Pattern Parade

Printed in the United States of America

ISBN 13: 978-0-15-360224-5
ISBN 10: 0-15-360224-4

1 2 3 4 5 6 7 8 9 10 039 16 15 14 13 12 11 10 09 08 07

Pattern Parade

by Joan Freese

Photographs by Kay McKinley

SCHOOL PUBLISHERS

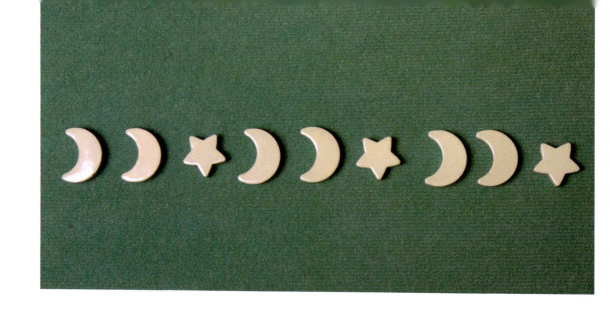

It is Pattern Week in Miss Penny's class. The class will learn about patterns. The children ask, "What is a pattern?"

"Let me show you," says Miss Penny.

triangle, square, circle, triangle, square, circle, triangle, square, circle

Miss Penny draws a blue triangle. Then she adds a blue square. Next she adds a blue circle. She repeats the three shapes to make a pattern.

purple, blue, purple, blue, purple, blue

Next, Miss Penny lines up balls.
Purple, blue, purple, blue. "What
comes next?" she asks.

"A purple ball," Toby answers.

"Then a blue ball," says Maya.

The colors make a pattern.

blue, yellow, blue, yellow, blue, yellow

The class really likes patterns. They want to make their own. Jack and Amelia make the same pattern with blocks. "Good job!" says Miss Penny.

blue, white, blue, white, blue, white

Miss Penny asks the class to look for patterns in other places. Simone says the floor tiles in the hall have a pattern.

big, small, big, small, big, small

Now it is time for recess. The class lines up near the wall. The children will play outside. They are holding big and small balls. The balls make a pattern.

1, 2, 3, 1, 2, 3, 1, 2, 3

Next Miss Penny shows the class another pattern. This pattern uses numbers. Miss Penny writes the pattern on the board. There are three numbers in this pattern unit.

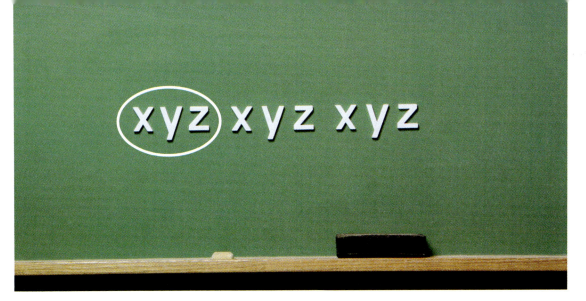

x, y, z, x, y, z, x, y, z

The class uses letters to make a pattern. Henry puts letters on the board. Miss Penny says, "Nice work! Your pattern unit is three letters long."

green, blue, yellow, green, blue, yellow, green, blue, yellow

The class looks for patterns in other places. Mimi says, "Jamie's top has stripes."

Miss Penny says, "Yes. That is a pattern too."

coat, hat, bag, coat, hat, bag, coat, hat, bag

Mark looks around for a pattern.
He does not look far! On the wall are
some hooks. He sees things hanging
there. The things make a pattern.

apple, apple, orange, apple, apple, orange, apple, apple, orange

The class is busy searching for patterns. Miss Penny shows the class some fruit. They see a pattern right away. The children tell Miss Penny what comes next.

white, white, black, white, white, black, white, white, black

The class makes the same pattern in the gym. They line up their right shoes. "Patterns are everywhere!" Juan says.

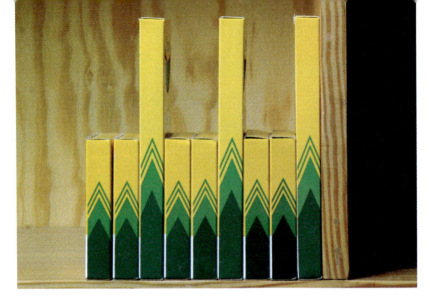

short, short, tall, short, short, tall, short, short, tall

Next, the class looks around for more patterns. Iris sees boxes on a shelf in the closet. The boxes make a pattern too.

pink, purple, purple, pink, purple, purple, pink, purple, purple

Lisa sees crayons in the art room. She has an idea. She forms a pattern with the crayons. Her friends help her pick what color comes next.

stand, stand, sit, stand, stand, sit, stand, stand, sit

Miss Penny has another pattern to share. "This pattern is three children long. Will you help me make it?" she asks.

"Yes," says the class. Everyone wants to help.

front, front, back, front, front, back, front, front, back

Then the class makes its own pattern. Everyone stands in a row. Two children face the front. One child faces the back. They repeat the pattern.

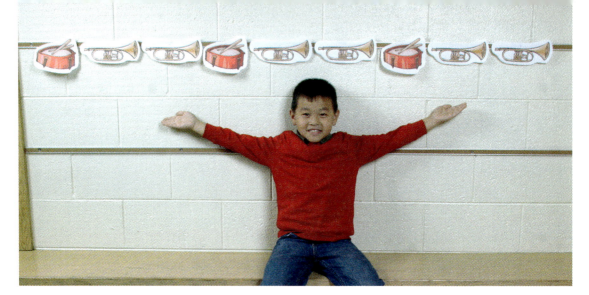

drum, horn, horn, drum, horn, horn, drum, horn, horn

Miss Penny's class is getting very good with patterns. Jake spots another pattern in the music room. It is on the wall.

white, blue, blue, white, blue, blue, white, blue, blue

Mimi has a pattern. "Look at my beads," she says. She shows them to the class.

It is sharing day in Miss Penny's class. The children bring things with patterns to school. The class has a pattern Show and Tell.

Then the class lines up. Miss Penny gives each child a crown to wear. It is Miss Penny's pattern parade!

Glossary

parade to march

repeat to do or say something again

tile a thin, hard plate used to cover a floor or roof